SpringerBriefs in Electrical and Computer Engineering

Cooperating Objects

For further volumes:
http://www.springer.com/series/10208

Pedro José Marrón · Daniel Minder
Stamatis Karnouskos

The Emerging Domain of Cooperating Objects

Definitions and Concepts

Springer

Pedro José Marrón
Networked Embedded Systems Group
University of Duisburg-Essen
Bismarckstr. 90
47057 Duisburg
Germany

Stamatis Karnouskos
SAP Research
SAP AG
Vincenz-Prießnitz-Str. 1
76131 Karlsruhe
Germany

Daniel Minder
Networked Embedded Systems Group
University of Duisburg-Essen
Bismarckstr. 90
47057 Duisburg
Germany

ISSN 2191-8112
e-ISSN 2191-8120
ISBN 978-3-642-28468-7
e-ISBN 978-3-642-28469-4
DOI 10.1007/978-3-642-28469-4
Springer Heidelberg New York Dordrecht London

Library of Congress Control Number: 2012933432

Printed on acid-free paper

Springer is part of Springer Science+Business Media (www.springer.com)

Preface

The book you have in your hands is the first of a series that will deal with all of the aspects related to Cooperating Objects, Internet of Things, Sensor Networks, Ubiquitous Computing, Cyber-Physical-Systems and Systems of Systems, just to name a few areas that are clearly related to each other. Although many of these topics have already been discussed in the scientific community for some time and not all of them are entirely new, there is the need to consolidate the current knowledge in these areas in order to jointly advance in our journey towards easy to use, widely deployed systems that, hopefully, make our lives easier through the use of technology.

The main purpose of the series is, therefore, to disseminate the current knowledge in these areas and address topics as diverse as algorithms, system solutions, applications, operating systems and even legal issues, as long as they are relevant to the topics at hand. The first books of the series have been mostly committed by members of CONET, the Cooperating Objects Network of Excellence, which is a European project funded by the European Commission to identify and produce seminal work on the main research topics in Cooperating Objects, thus shaping the academic and industrial research in the short, medium and long-term. To what extent this has been achieved by the project itself is left to the discretion of the reader.

The current issue tries to shed some light into the different concepts, terminology and lines of research currently being discussed in the context of Cooperating Objects. Having a clear definition of concepts is crucial to have a common base to discuss further topics, and also serves the purpose of setting up the Springer Brief Series in a well-defined frame. We have also tried to put some boundaries to research areas that, with different names, seem to concentrate on very similar things. These areas, mentioned above, have grown organically within the different communities but converge more and more with the increasing importance of communicating and cooperating entities in the last years. In the most cases, it is not easy to find these boundaries and, depending on the definition used, might fall on the one or the other side of the spectrum. Nevertheless, having a pictorial representation of the different research areas and how they relate to each other based on the current definitions is a valuable contribution that should be revised and refined regularly.

Finally, it remains only to hope that you enjoy not only this first book but all following in the series and that you find them informative and interesting.

Duisburg, Karlsruhe
December 2011

Pedro José Marrón
Daniel Minder
Stamatis Karnouskos

Acknowledgments

This book has been partially supported by CONET, the Cooperating Objects Network of Excellence (www.cooperating-objects.eu), funded by the European Commission under FP7 with contract number FP7-2007-2-224053.

We would also like to express our gratitude to all of the members of the CONET consortium that have contributed greatly with their knowledge, insight in the topics and discussions to improve this book.

Contents

Chapter 1
Foundation

The world is converging in many different ways and coming closer together, also at the social level, mainly through the use of technology. Ten years ago nobody would have been able to predict the current market share of smart phones (over one billion activated smart phones in 2011), or the prevalence of highly complex ICT solutions in cars and electronic appliances used in our everyday lives. If we turn to research, the advances in the last years have been astonishing in a similar way. Research areas that were traditionally separated are finding more and more commonalities and coming together with the combination of interdisciplinary skills focusing on a common problem.

This trend is exciting since it allows for the creation of new areas of research as well as the analysis of existing technologies from different angles, enabling the creation of surprising solutions and offering new insights that shape the future in unprecedented ways. In the past years, it has become apparent that interdisciplinarity and the ability to work in heterogeneous teams composed of researchers, industrial partners and end-users coming from different but complementary areas of expertise, have become crucial skills needed to tackle the current technological challenges.

The area of Cooperating Objects is no exception to this trend and has been defined in the past years by a team of experts at the European level as the combination of different types of systems that have become more and more relevant recently in the broader context of embedded systems. First, there is the classic concept of **embedded systems**, defined mainly as a control system for physical process (machinery, automobiles, etc.). More recently, the notion of pervasive and **ubiquitous computing** started to evolve, where objects of everyday use can be endowed with computational capacity, and perhaps with simple sensing and communication capabilities. Later on, the notion of **Wireless Sensor Networks** appeared, where entities that sense their environment not only operate individually, but cooperate using ad-hoc network technologies to achieve a well-defined purpose of supervision or monitoring of an area, a particular process, etc.

Our claim is that these three types of systems (i.e. embedded systems, pervasive and ubiquitous computing and wireless sensor networks), that act and react on their environment, are quite diverse and novel systems on the one hand, but also

P. J. Marrón et al., *The Emerging Domain of Cooperating Objects*,
SpringerBriefs in Cooperating Objects, DOI: 10.1007/978-3-642-28469-4_1,

share fundamental commonalities at the same time. For this reason, there is a natural complementarity to their capabilities that allows them to interact with their environment and cooperate to achieve a common goal. In particular, important notions such as control, heterogeneity, wireless communication capabilities, dynamics/ad-hoc nature, and cost are present to various degrees in each one of these types of systems.

Let us now briefly outline each one of these areas in more detail and provide some references to in-depth literature.

1.1 Embedded Systems

The term embedded system refers to a microprocessor-based system that is part of ("embedded") a larger system and designed for specific control, monitoring or data-processing functions. The general characteristics of an embedded system are varied and greatly depend on the task that needs to be solved. In general, most embedded systems have some form of real-time constraints that they have been designed to meet, although this is not a must. Since embedded systems are specifically designed both in hardware and software for a specific task, they can be optimised to reduce the size and the cost of the product, while at the same time increasing its reliability, robustness and performance [1–3].

The history of embedded systems dates back to the Apollo Guidance Computer, which was built from 4.100 integrated circuits and was used first in 1965. The appearance of the first commercially available single-chip microprocessor in 1971, the Intel 4004, and some years later the development of single-chip microcontrollers, which combined processor code, memory, and I/O ports, such as Texas Instrument's TMS 1000 in 1974 boosted the adoption of embedded systems.

Today, embedded systems already span all aspects of modern life and have been widely used in mobile phones, digital cameras, DVD players, household appliances, ATM machines and very extensively in cars, just to name a few. The trend is towards even more embedded devices and, if we look at the evolution of the automobile industry in the last years, for example, the usage of microprocessors and microcontrollers in cars has increased from about a dozen to well over 100, thus increasing both, the complexity of the car itself but also its reliability and ability to adapt to very different driving styles and conditions.

From the point of view of hardware, embedded systems use a variety of microprocessors and microcrontrollers. ARM [4], AVR [5] and MIPS-based [6] architectures are very popular, but also processors based on Intel's x86 platform are included in many products. For very high-volume embedded systems ASICs or FPGAs can be found. However, the most important characteristic of embedded systems is that they operate efficiently both in terms of resource usage (such as energy) and in terms of software, which has lead to the creation of special purpose operating systems such as QNX [7] or VxWorks [8] that run on high-end embedded systems, although MS-DOS and Linux are also common.

1.2 Ubiquitous Computing

The term ubiquitous computing was coined by the late Mark Weiser in his seminal paper [9], where he described the fact that the most astounding technologies were the ones that disappear. Since that date, Ubiquitous Computing, also called Pervasive Computing, has dealt with the creation of the technologies needed to include computers in all of our everyday activities in a completely transparent and seamless way.

For this reason, Ubiquitous Computing is not a particular discipline but encompasses technologies that deal with very different aspects such as Human Computing Interaction, Wearable Computing, Augmented Reality, Context-Aware Systems and many other. However, in all of these disciplines, the ultimate goal is to enable the user to perform new and improved interactions with the environment. Equally important to technology and computer science issues are the psychological and social aspects of people interacting with technology.

At its technical core, Ubiquitous Computing uses small, inexpensive and robust devices integrated into everyday objects. These devices can form networks using inexpensive short-range wireless radios. A central concept of Ubiquitous Computing is context-awareness. The devices perceive the environment and deduce the situation of persons, locations or objects. Based on this information, the networked system selects appropriate data or services that are presented to the user, changes its way of presentation, or performs other actions. As a result, the interaction with this Ubiquitous Computing devices become intuitive since it does not rely on traditional human-computer-interfaces like keyboard, mouse or screen.

Please, refer to [10] for a good introduction to the complexity of the field and a good summary of current technologies used to support the vision of truly ubiquitous computing.

1.3 Wireless Sensor Networks

Sensor networks are composed of a collection of small devices that are capable of sensing, processing data, and communicating this information to other neighbouring devices mostly using wireless technology. Recent advances in communication and hardware technologies have allowed the continuous miniaturisation of such devices so that, in the not so distant future, sensor networks might be pervasive and allow us to interact with our environment in much richer ways.

Although in computer science the paper "Next Century Challenges: Scalable Coordination in Sensor Networks" from 1999 by Deborah Estrin et al. [11] has definitely boosted the amount of attention given to sensor networks, research in this field (especially in the U.S.A.) has a long history [12].

Modern research on sensor networks started as a way to improve military applications and with the development of the Distributed Sensor Networks (DSN) program at the Defense Advanced Research Projects Agency (DARPA) around 1980. At a time where the Internet consisted of about 200 hosts and TCP/IP had just been designed, the first *Distributed Sensor Nets Workshop* held at Carnegie Mellon University in 1978 contributed to identify the technology components needed for a distributed sensor network. These were the following:

- Acoustic sensors.
- High level protocols to allow for the communication between the different components.
- Common processing techniques and algorithms, including self-localisation.
- Distributed software techniques, including dynamically modifiable distributed systems and language design.

Although the hardware back then was primarily composed of PDP-11 and VAX machines running Unix and VMS, the requirements and challenges were very similar to the ones defined today as the characteristics and hard research challenges of sensor networks.

Interestingly enough, the hardware has simply become smaller, but its characteristics in terms of processing power, working memory and program memory sizes remain similar to the computers found in the early 1980s, which memory sizes between 64 and 128 kBytes and memory sizes of up to 8 kBytes.

Akyildiz et al. [13] consider such resource limitations, especially the limited amount of power available to sensor nodes, to be one of the intrinsic characteristics of sensor networks. They compare sensor networks to ad-hoc networks and point out the following differences that, in essence, summarise the features of sensor networks:

- The number of sensor nodes can be several orders of magnitude higher than the number of nodes in an ad-hoc network.
- Depending on the type of application, sensor networks can be very densely deployed so that short-range, broadcast-based and multi-hop communication plays a very important role.
- Sensor nodes are prone to failures, especially in the case where sensor nodes are used to monitor outside events.
- Sensor nodes are very limited in power, computational capabilities and memory.

Nowadays, applications of sensor networks have the following characteristics:

- Communication takes place in multi-hop fashion only if it consumes less energy than direct communication to other nodes (for example, a base station).
- Hybrid network topologies are starting to become relevant, especially in areas where the nature of the problem requires some base station, or existing infrastructure simply cannot be ignored.
- Although sensor networks have been traditionally stationary, current trends indicate that more and more applications need the capability of mobile sensor nodes, which require algorithms that are fundamentally different from their static counterparts.

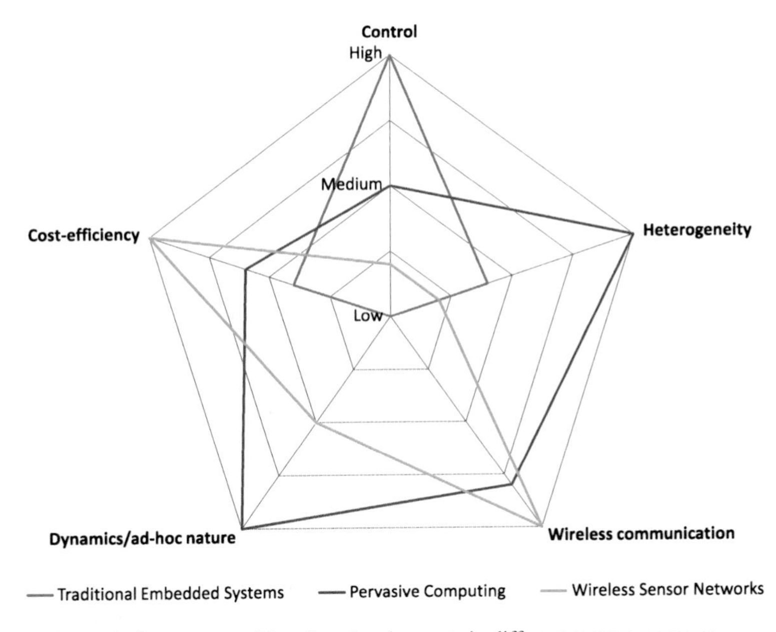

Fig. 1.1 Strengths/Importance of key functional aspects in different system concepts

- Finally, new application domains imply the use of sensors and actuators in the same network, giving rise to the concept of *Sensor Actuator Networks*, where challenges regarding timeliness of signals, coordination among actuators, etc. play an important role.

Regarding general purpose books, there are a number of books that can be read that deal with all of the aspects mentioned above [14–16].

1.4 Convergence Criteria

Figure 1.1 shows the different weights of these functional aspects. As already mentioned, the strength of traditional embedded systems is their control functionality. Pervasive computing applications include control aspects as well, but usually do not have hard real-time constraints. Heterogeneity is a key aspect of pervasive computing since no common platforms can be assumed given the potential breadth of mass market solutions. In contrast, single embedded systems and wireless sensor networks usually operate in a controlled setting where heterogeneity is typically low. However, if several embedded systems are combined as, for example in a car, their heterogeneity increases. A characteristic of wireless sensor networks and pervasive computing is wireless communication whereas in contrast traditional embedded systems are wired. This has direct implications on dynamics and ad-hoc nature since wired systems are static. The dynamic nature of pervasive computing is tightly related with its heterogeneity. For wireless sensor networks there exist both static and mobile scenarios. Since many of them assume a high number of sensor nodes low cost is very important. If many devices of our environment should be integrated in pervasive computing applications the cost aspect will become more relevant especially for cheap devices. Since embedded systems are usually integrated into larger and more expensive devices the cost of the single embedded system is less important than for the other system concepts.

The conception of a future-proof system would have to combine the strong points of all three system concepts at least in the following functional aspects:

- Support the control of physical processes in a similar way as to what embedded systems are able to do today.
- Have as good support for device heterogeneity and spontaneity of usage as pervasive and ubiquitous computing approaches have today.
- Be as cost efficient and versatile in terms of the use of wireless technology as Wireless Sensor Networks are.

The convergence of these three types of technologies that, until now, have been evolving independently of each other (Fig. 1.1), is what we call Cooperating Objects technologies. This new term is born out of the combination of these traditional systems.

Moreover, this notion or paradigm of Cooperating Objects is even stronger than the individual technologies it stems from, as it carries over to their internal structure, highlighting the diversity of cooperating patterns admissible under this general paradigm. Also, pointing to the importance of complementing the vision of pervasive computing with that of pervasive control is essential.

In view of the emergence of new technologies and devices, their increasing integration into the everyday life and the need to coordinate them with a view to making communication easier mainly as to the interoperability, mobility and the scalability, Cooperating Objects are regarded as a key enabler and aim at providing a proactive support to users or machines in their collaborative tasks. Indeed, the major advantage of the Cooperating Object lies in the possibility to tackle the complexity of the new surrounding environments due to the high number of involved devices or systems and the heterogeneity of components. As depicted in Fig. 1.2, we see a paradigm shift where cooperation will be the key aspect in the next generation of networked embedded systems that will have to function in a highly heterogeneous environment composed of virtual and real-world devices.

The generality of this model allows us to seamlessly include different fields like sensor networks, pervasive computing, embedded systems, etc. As an example, consider the following scenario: nowadays, we have at our disposal lots of information, data sources or systems or even services, like the traffic panels along the main roads, traffic radio, GPS devices or web services to plan a trip. GPS devices may give alternatives but only under the human initiative and not all commercial GPS take into account real-time traffic data yet. However, video cameras and other surveillance systems are able to provide some of this data. And if we integrate the Traffic

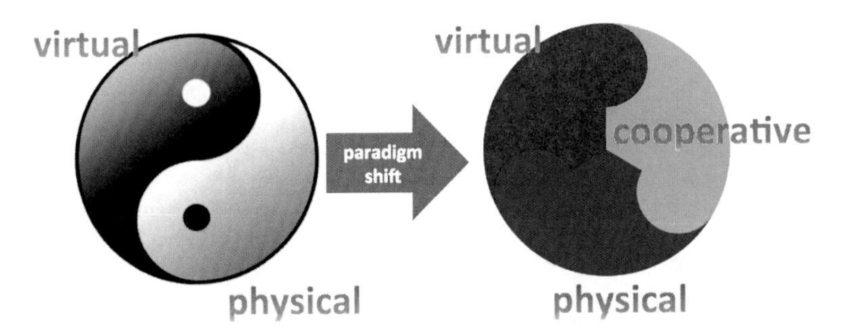

Fig. 1.2 Cooperation as the key characteristic of future networked embedded systems

Message Channel services (TMC technology) for example, we get another useful flow of information which could be computed. To be able to achieve our goal, a proactive process is necessary, and for this reason the cooperation between all of these various information sources is crucial. How to reach that, i.e. a cooperative surrounding environment to link vehicle and infrastructures? One of the solutions would be to use Cooperating Objects, which would make it possible to drop some current barriers between these elements such as heterogeneity, complexity, scalability and to improve the communication with a view to providing ad-hoc networks, thus data mobility would be enhanced. In this scenario, the Cooperating Object might consist of several parts: for instance one that continuously measures local traffic data, a second one to integrate all traffic-related data from available information flows from infrastructures and another that asks the GPS device in view to offering route alternatives.

As can be seen, the rapid advances in computational and communication in embedded systems, are paving the way towards highly sophisticated networked devices that will be able to carry out a variety of tasks not in a standalone mode, as usually done today, but taking into account dynamic and context specific information. These "objects" will be able to cooperate, share information, act as part of communities and generally be active elements of a more complex system.

References

1. Noergaard T (2012) Embedded systems architecture: a comprehensive guide for engineers and programmers. Newnes, Oxford
2. Simon D (1999) An embedded software primer. Addison Wesley, Boston
3. Heath S (2003) Embedded systems design. Newnes, Oxford
4. Jaggar D (1997) ARM architecture reference manual. Prentice Hall, London
5. Gadre DV (2000) Programming and customizing the AVR microcontroller. Mcgraw-Hill Professional, New York
6. Heinrich J (1991) MIPS RISC architecture. Prentice Hall, Englewood Cliffs
7. Kolwick (1989) Qnx operating system. Basis Computer Systems
8. Wehner C (2006) Tornado and VxWorks: what's not in the manual. Books on Demand GmbH.
9. Weiser M (1991) The computer for the 21st century. Sci Am 43(3):66–75

10. Poslad S (2009) Ubiquitous computing: smart devices, environments and interactions. Wiley, London
11. Estrin D, Govindan R, Heidemann J, Kumar S (1999) Next century challenges: scalable coordination in sensor networks. In: Proceedings of the 5th annual ACM/IEEE international conference on mobile computing and networking, ACM Press, pp 263–270. doi:10.1145/313451.313556
12. Chong CY, Kumar SP (2003) Sensor networks: evolution, opportunities, and challenges. Proc IEEE 91(8):1247–1256
13. Akyildiz IF, Su W, Sankarasubramaniam Y, Cayirci E (2002) A survey on sensor networks. IEEE Commun Mag 40(8):102–114
14. Karl H, Willig A (2005) Protocols and architectures for wireless sensor networks. Wiley, Berlin
15. Callaway EH (2003) Wireless sensor networks—architectures and protocols. Auerbach, Boca Raton
16. Zhao F, Guibas L (2004) Wireless sensor networks—an information processing approach. Elsevier/Morgan-Kaufman, Amsterdam

Chapter 2
Definition of Cooperating Objects

The term "Cooperating Objects" was coined by the Embedded WiSeNts Research Roadmap [1] in November 2006. The Cooperating Objects Network of Excellence CONET used the same definition in the "Research Roadmap on Cooperating Objects" [2]. For the second edition of the roadmap "The emerging domain of Cooperating Objects" [3] the definition was revised to emphasise the cooperative aspects. We develop this definition further to improve its clarity as follows:

> Cooperating Objects are modular systems of autonomous, heterogeneous devices pursuing a common goal by cooperation in computations and in sensing and/or actuating with the environment.

As explained in the previous chapter, the domain of Cooperating Objects is a cross-section between (networked) embedded systems, ubiquitous computing and (wireless) sensor networks. There are, therefore, several flavours of Cooperating Objects depending on the degree in which they fulfill different Cooperating Object features. Some of them can process the context of cooperation intentionally, act on it and intentionally extend it, change it or stop it. As such they may possess the necessary logic to understand semantics and build complex behaviours, thus allowing the Cooperating Object to be part of a dynamic complex ecosystem.

As depicted in Fig. 2.1, there are several areas that share common ground with Cooperating Objects e.g. software agents, Internet of Things, Cyber-Physical systems etc. However what differentiates them is the mix of the degree of physical and feature elements that creates the right recipe for a specific area. For instance Cooperating Objects focuses mostly on the cooperation aspects while considering the rest of them only as enabling factors to achieve cooperation. Other approaches e.g. Internet of Things, focus mostly on the interaction and integration part while cooperation is optional. So the differentiating factor among all areas, is not the distinct characteristics but which of them they employ (depending on the scenario) and at which degree.

In the following sections each of the key features of this definition is discussed in detail.

P. J. Marrón et al., *The Emerging Domain of Cooperating Objects*,
SpringerBriefs in Cooperating Objects, DOI: 10.1007/978-3-642-28469-4_2,

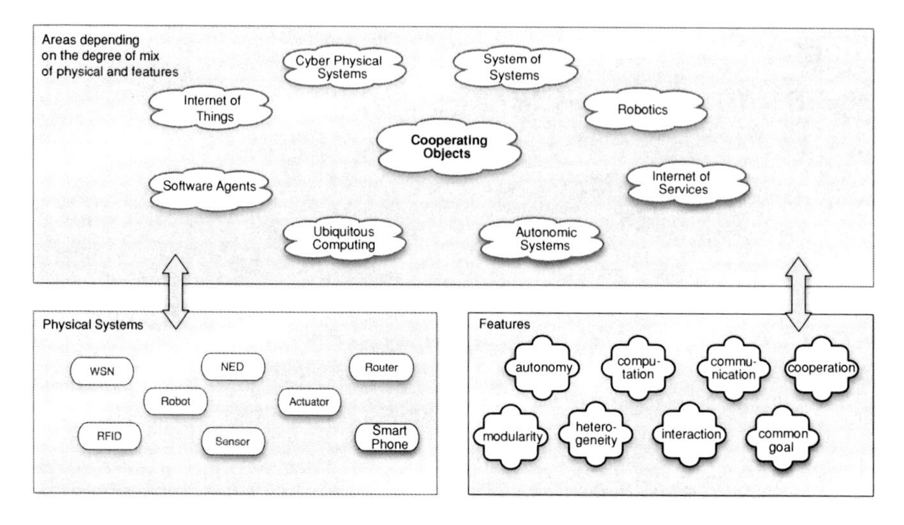

Fig. 2.1 Overview of various areas

2.1 Modularity

A Cooperating Object is composed of several devices that need to exhibit certain features according to the definition. Each of the devices contributes functionality to the overall Cooperating Object, but the modularisation helps to keep the single devices simple and maintainable.

The logical organisation of the single devices inside the Cooperating Object is not restricted by the definition and can range from completely flat to completely hierarchical. In the latter case, another Cooperating Object is included completely in a superordinate Cooperating Object that contains more devices or sibling Cooperating Objects. From the top-level Cooperating Object all participating devices and Cooperating Objects have a similar appearance since there is usually a single device representing the subordinate Cooperating Object to the superordinate one. Thus, in all following explanations a device can also be a complete Cooperating Object unless stated otherwise. The use of hierarchies depends on the type of cooperation but also on the need for scalability for which hierarchies are a proven principle in distributed systems.

Single devices can be exchanged if the replacement provides the same functionality. Thus, modularity is a prerequisite in a dynamic environment where the availability of all devices belonging to the current configuration cannot be guaranteed at all times. In contrast, it is quite common that devices are replaced by other devices during the lifetime of a Cooperating Object.

The modular design makes it also possible to replace a device by a more powerful one or to add new devices that extend the functionality. Thus, the Cooperating Object can be developed in an evolutionary fashion and adapted to new needs.

In contrast, a single device cannot be a Cooperating Object although it might consist of several components. In these monolithic systems the single components are not independent (see Autonomy in Sect. 2.2) and their dynamic exchange is impossible.

As for every modular system, the design of the interface and APIs is a crucial point. In general, the interface should protect the device itself while allowing others to use the provided functionality in a controlled way. Interfaces should adhere to standards if they exist so that the possibility for other matching modules is higher. They should not be tailored to a specific device since this creates a tightly coupled system which might only work in exactly one configuration, leading to an almost monolithic system.

2.2 Autonomy

Each device can decide on its own about its involvement in a Cooperating Object. If the Cooperating Object does not participate at all in the cooperation and coordination activities, it is not considered part of the Cooperating Object. Otherwise, it decides about the degree of participation. In general, a Cooperating Object can dedicate only a fraction of its resources or its functionality to the current Cooperating Object, thus leaving the possibility to serve multiple Cooperating Objects. Note that this is not restricted by the definition.

In contrast, the device can also exclusively "belong" to a single Cooperating Object. This is not a contradiction to the modularity principle as long as the device is not directed to another particular device, for example if they are statically wired, but can decide autonomously with which device it is cooperating. Nevertheless, the device might be designed for a special functionality so that it can only be able to participate with a special type of devices. However, it is still the autonomy of the device to choose with which particular instance of them.

The decision to cooperate or not can be based on various factors, e.g. the current and maximum number of Cooperating Objects a device is able to join, the energy level of the device, past interactions with the other devices requesting cooperation, compatibility of goals, and various application-specific variables. The single factors can be combined in a single cost function to evaluate the willingness to cooperate in an easier fashion. Different flavours of this willingness can exist: a device may simply be providing its information to all other devices that want to have it, i.e. this "helper device" always cooperates. On the other hand, a device can evaluate if it will benefit from cooperating, for example because the other devices could provide functionality in turn that it needs for its own tasks. Other incentive mechanisms can improve this mutual willingness to cooperate. Another possibility is that the requested task is compatible to an already executed task on the device and, thus, the cooperative work can be performed without or with only very little additional effort.

Autonomy also implies that there is no master that assigns the membership to a Cooperating Object or the functionality inside it. However, it is possible that the devices of a Cooperating Object autonomously select a (temporary) master that configures the Cooperating Object. Usually, this master property circulates over time through the devices.

2.3 Heterogeneity

In the definition of Cooperating Objects, heterogeneity is a crucial point since it is more than heterogeneity in terms of, e.g. processing power or memory. In fact, a Cooperating Object must combine devices of different system concepts, i.e. Wireless Sensor Networks, embedded systems, robotics, etc. Since devices belonging to these different concepts often have different hardware characteristics the heterogeneity is also exhibited in this respect, but it is a consequence and not the actual meaning of "heterogeneity" in this definition.

Taking into account this meaning, it is obvious that a Cooperating Object is more than one of the system concepts presented in Chap. 1 that constitute the basis of Cooperating Objects. The novelty and at the same time the challenges of Cooperating Objects arise from the demand to strengthen the underdeveloped functional aspects of one the system concepts by combining it with other concepts without loosing the strengths.

Of course, the research challenges of the separate system concepts do not disappear merely by combining them, but they should be tackled in their areas. Cooperating Objects research should instead concentrate on the new challenges arising from the combination. However, there will be an overlap since problems can be equal for Cooperating Objects and an underlying system concept.

2.4 Computation

The majority of Cooperating Objects are computing devices cooperating with each other. (Electro-)Mechanical parts can be connected to these devices but do not count as extra devices (see Autonomy in Sect. 2.2). Due to the different nature of the single devices in a Cooperating Object (see Heterogeneity in Sect. 2.3) the computational capabilities can vary largely. However, a device must at least be able to take an autonomous decision about its involvement in a Cooperating Object and to communicate with other devices, which usually requires also computation.

Local computation can also be used to reduce the need for or the amount of communication, e.g. by compression or filtering, which can be a goal in energy-restricted scenarios. On the other hand, Cooperating Objects with less powerful devices have to distribute the computational workload onto several devices, which is an important self-configuration task.

2.5 Interaction with Environment

Cooperating Objects interact with the environment using sensors and/or actuators. Sensors convert physical or chemical quantities into a signal which is read by the connected computational device, and actuators transform a signal from this device into mechanical motion or physical quantities. It strongly depends on the application which sensors and actuators are involved. It can be as simple as a thermistor or an LED, but can also e.g. consist of an external digital signal processor or a machine controller. Nevertheless, the interaction should be substantial, especially with respect to actuators, i.e. actuation should have a changing effect on the environment.

The involvement of sensors and actuators makes Cooperating Objects real-world objects, i.e. there are no pure virtual Cooperating Objects. Certainly, not every device needs to have sensors or actuators, i.e. there can be computation-only devices, but sensing and/or actuation must be performed by some devices of the Cooperating Objects and this interaction with the environment must be a core functionality of the Cooperating Object and not just an optional side-effect.

2.6 Communication

Interestingly, the first of the five pragmatic axioms on human communication that Paul Watzlawick formulated in 1967 is also valid for Cooperating Objects: "one cannot not communicate" [4]. If there is no observable behaviour of a device at all, it implicitly denies all cooperation—be it by an autonomous decision or simply because the device failed. After all, communication is a requirement for cooperation, although not explicitly mentioned in the definition of Cooperating Objects.

If a device communicates there are three techniques of information exchange [5]: the most obvious technique is explicit communication, which can be performed using various means, e.g. wires, radio, light, sound. The content of the communication is manifold and can range from just the state of the single device to a common planning. Cooperating Objects do not restrict the possible communication patterns since it might still be necessary to communicate with a single device, all devices or a certain group of devices.

Besides explicit communication, there are two other techniques that work by observation using sensors. With passive action recognition the actions of other devices are observed, e.g. if an actuator moves. In contrast, the effects of actions of others can be sensed ("stigmergy"), e.g. the increase of temperature caused by a heater. Usually, these forms of communication show the lack of common interfaces for direct communication; nevertheless, the inclusion of such devices allows for interesting applications.

2.7 Common Goal

The ultimate reason for a Cooperating Object to exist is the common goal it tries to achieve. There should be a reason for pursuing the goal using Cooperating Objects: either the goal can only be achieved through Cooperating Objects or there is at least an improvement compared to a monolithic or centralised approach.

The form of goal definition is not restricted. Automated planning methods could be applied when the initial state is derived from the context and the possible actions are gathered from all available devices. However, due to the complexity of these planning algorithms, the goal is normally not explicitly stated at all, but represented as an application that describes also the required external interfaces. By finding devices that run the application and provide the needed interfaces the Cooperating Object is built up and the tasks are distributed among the devices. This task distribution has to take into account that some devices can execute certain tasks at all or in a more efficient way than other devices due to heterogeneity. However, it is the eventual result of the autonomous decision of each device.

If a device is only the delegate for a subordinate Cooperating Object the described planning and/or application building process for this task resulting in sub-tasks for the devices that are involved in the subordinate Cooperating Object.

Although the devices do not know the overall goal they execute a task to achieve it. Thus, each device has detailed knowledge only about its area of responsibility, but limited information about the whole Cooperating Object. However, the cooperation of the single devices make it possible to achieve the overall goal, which needs the full picture. Thus, the intelligence of the system lies distributed in the network.

Due to the potential long-running nature of Cooperating Object applications it is obvious that goals will change over time. This results in different tasks and, thus, in a reconfiguration of the Cooperating Object.

2.8 Cooperation

"Cooperation" is a widely and often used term, but its meaning differs between areas; and also inside an area there is not necessarily a common agreement. Moreover, the terms "cooperation" and "collaboration" are often used as synonyms. For example, in the Merriam Webster thesaurus [6] both words have the same definition: "the work and activity of a number of persons who individually contribute toward the efficiency of the whole". Therefore, we review some of the definitions before presenting our intention of "cooperation".

The Macmillan Dictionary [7] defines cooperation as "a situation in which people or organisations work together to achieve a result that will benefit all of them" and collaboration as "the process of working with someone to produce something". It can be argued that according to these definitions cooperation has a stronger motivation for common work than collaboration, but this does not hold for most other definitions.

In [8] both terms are defined with respect to agents. For cooperation, agents have defined roles that cannot be changed and, therefore, they must cooperate. The roles are assigned by the designer with a (single) goal in mind. In contrast, dynamic links between agents without predefined roles lead to collaboration. This way, agents can achieve their own goals by negotiating contracts with other agents. If there is a common goal both agents contribute shared effort to this goal. If no common goal can be negotiated but the single goals are well aligned, only resources can be shared in such a way that both agents can benefit. For Cooperating Object we do not adopt this distinction based on the difference between static and dynamic roles.

Three important categories of interactions between (computing) processes are described in [9]: cooperation is anticipated and desired interaction; competition is anticipated and acceptable, but undesirable interaction; and interference is unanticipated or unacceptable. Competition usually happens when access to computing resources needs to be serialised. With real-world objects, competition for resources of the real world occurs as well, e.g. cars driving on a crossing. Interference happens when the actions of unrelated devices collide and reduce the fulfilment of the goal. However, when combining the devices in a Cooperating Object they will coordinate before competition or interference can happen at all, thus making the overall process more efficient. It is also known from economic science that cooperation allows a more efficient use of resources than competition [10].[1]

According to [11], which deals with computer-based learning environments, "[c]ooperative work is accomplished by the division of labour among participants, as an activity where each person is responsible for a portion of the problem solving", i.e. the task is split in independent sub-tasks that can be performed individually. In contrast, collaboration needs "[...] the mutual engagement of participants in a coordinated effort to solve the problem together", i.e. there is no or only partial division of the work but a continued combined activity.

A similar direction is taken by [12], which establishes a hierarchy between coordination, cooperation and collaboration—we only consider the latter two. For cooperation, a mutual benefit should be gained, e.g. savings in time and cost, by sharing or partitioning work. For collaboration, a collective result should be achieved, which would not be possible alone. This result should be innovative, extraordinary or a break-through. A typical collaborative task would be brainstorming, i.e. a creative process.

For Cooperating Objects, we follow this view. According to our definition at the beginning of this chapter, there is a common goal that the devices try to achieve. As described in Sect. 2.7 the devices finally execute individual tasks. Nevertheless, some of these tasks might also need tight coordination due to the limitation of the participants or, for example, when two actuators driven by separate device need to interact. However, this is not comparable to an ongoing, creative, problem solving task as described by the last two definitions.

[1] Cooperation and competition can also be seen as orthogonal instead of opposed. Both span a continuum from weak to strong, and a relationship between two firms can be placed anywhere on this matrix. This phenomenon is called coopetition.

In our sense, cooperation is always intentional and driven by a goal. Without a goal and, thus, no tasks there is no need for cooperation at all. Although unintentional interaction might deliver the same results it does not happen in a controlled way which creates problems in case of errors. For example, reconfiguration is more difficult if the exact task that a device has performed is not known.

The participation of all devices in a Cooperating Object is needed to achieve the common goal, i.e. a Cooperating Object is more than just the sum of the single devices. Nevertheless, the common goal (see Sect. 2.7) does not imply benefits for all the cooperating devices. Some of them can be especially designed to help in cooperation, others can play a more active part in one cooperative task to profit more in another one. When autonomous and selfish objects decide autonomously if and how they cooperate the sum of the benefit must be positive. Otherwise, a device will eventually not agree to cooperate or not be asked to cooperate any more.

References

1. Marrón PJ, Minder D (eds) (2006) Embedded WiSeNts consortium, Embedded WiSeNts research roadmap. Logos, Berlin
2. Marrón PJ, Karnouskos S, Minder D (eds) (2009) Research roadmap on cooperating objects. European Commission, ISBN: 978-92-79-12046-6. Office for Official Publications of the European Communities. doi:10.2759/11566
3. Marrón PJ, Karnouskos S, Minder D, Ollero A (eds) (2011) The emerging domain of cooperating objects. Springer, Berlin. http://www.springer.com/engineering/signals/book/978-3-642-16945-8
4. Watzlawick P, Beavin JH, Jackson DD (1967) Pragmatics of human communication. W. W. Norton, New York
5. Siciliano B, Khatib O (eds) (2008) Springer handbook of robotics. Springer, Berlin
6. Merriam-Webster (2011) Merriam-Webster thesaurus. http://www.merriam-webster.com/thesaurus/
7. Macmillian (2011) Macmillan dictionary. http://www.macmillandictionary.com/
8. Sioutis C, Tweedale J (2006) Agent cooperation and collaboration. In: Gabrys B, Howlett R, Jain L (eds) Knowledge-based intelligent information and engineering systems. Lecture notes in computer science, vol 4252. Springer, Berlin, pp 464–471. doi:10.1007/11893004_60
9. Horning JJ, Randell B (1973) Process structuring. ACM Comput Surv 5:5–30. doi:10.1145/356612.356614
10. Kohn A (1992) No contest: the case against competition. Houghton Mifflin, New York
11. Roschelle J, Teasley SD (1995) The construction of shared knowledge in collaborative problem solving. In: O'Malley C (ed) Computer-supported collaborative learning. Springer, Berlin, pp 69–97
12. Ditkoff M, Moore T, Allen C, Pollard D (2005) The ideal collaborative team. http://www.ideachampions.com/downloads/collaborationresults.pdf

Chapter 3
Related Concepts

As mentioned before, Cooperating Objects is an emerging domain that can be identified by the key characteristics of the participating devices and systems, predominantly from the cooperation aspects of the constituted objects that depend both on the virtual as well as real world. Clearly they represent an evolutionary step of pre-existing approaches and built upon them. There are several concepts that share common ground with Cooperating Objects such as Cyber-Physical systems, Internet of Things, Internet of Services, M2M, robotics, system of systems, autonomic systems, etc. In this chapter we focus on the definition of each related area and show the similarities and differences with Cooperating Objects. Additionally we give some examples of domains where Cooperating Objectsplay a pivotal role in order to better make understandable the context they operate on.

3.1 Technologies and Concepts

3.1.1 Cyber-Physical Systems

In 1996, the US National Science Foundation (NSF) launched a research initiative on Cyber-Physical Systems, starting with a workshop in Austin, Texas, to "usher in a new generation of engineered systems that are highly dependable, efficiently produced, and capable of advanced performance in information, computation, communication, and control" [1].

Since September 2008 the NSF has issued three program solicitations on the topic "Cyber-Physical Systems", namely "NSF 08-611", "NSF 10-515" and "NSF 11-516". First projects in this program started on Sep 1st, 2009. To date (Oct 2011) 65 projects have been funded with a total grant of more than $57 Mio.

P. J. Marrón et al., *The Emerging Domain of Cooperating Objects*,
SpringerBriefs in Cooperating Objects, DOI: 10.1007/978-3-642-28469-4_3,

The definition of Cyber-Physical Systems (CPS) has not been changed between the single solicitations. In NSF 11-516 [2] it is given as follows:

> The term 'cyber-physical systems' refers to the tight conjoining of and coordination between computational and physical resources. [...] These capabilities will be realised by deeply embedding computational intelligence, communication, control, and new mechanisms for sensing, actuation, and adaptation into physical systems with active and reconfigurable components.

The text lists several areas in which tomorrow's CPS should outmatch today's systems: "adaptability, autonomy, efficiency, functionality, reliability, safety, and usability". Compared to the older solicitations, the newest explicitly lists "fault tolerance, availability, reliability, reconfigurability, and cyber-security aspects of certifiably-dependable CPS" as important concepts that should be tackled.

Several possible application areas are mentioned in the text, e.g. autonomous collision avoidance, robotic surgery, fire-fighting, automated traffic control, zero-net energy buildings, ubiquitous healthcare monitoring. The latest solicitation especially encourages "novel CPS manufacturing approaches [...]; CPS technologies to enable sustainable and dependable energy and water resources [...]; next generation automotive, aerospace, and rail CPS [[...]]; new concepts for medical CPS and healthcare delivery CPS; CPS for sustainable agriculture."

The definition and the applications are mostly in line with Cooperating Objects. While Cyber-Physical Systems stress more the relation between computational and physical elements of the system, Cooperating Objects put more emphasis on the cooperation aspect of the single objects. However, the research areas and concepts that emerge from both definitions are similar.

Apart from the NSF definition, other groups have presented their view on Cyber-Physical Systems as well, which disturbs this clear picture. For example, the Steering Group of the CPS Summit 2008 provided a more precise definition in its "Cyber-Physical Systems Executive Summary" [3]:

> A CPS is a system:
>
> - in which computation/information processing and physical processes are so tightly integrated that it is not possible to identify whether behavioural attributes are the result of computations (computer programs), physical laws, or both working together;
> - where functionality and salient system characteristics are emerging through the interaction of physical and computational objects;
> - in which computers, networks, devices and their environments in which they are embedded have interacting physical properties, consume resources, and contribute to the overall system behaviour.

The report states that the goal is "to add capabilities to physical systems that we could not feasibly add in any other way" and illustrates that CPS "range from minuscule (pace makers) to large-scale (the national power-grid)".

Here, the cyber-physical integration is the most prominent part which already makes a single electromechanical object that is tightly integrated with the physical world (pacemaker) a Cyber-Physical System. In contrast, we will not consider this example as a Cooperating Object since the cooperative part of several devices interacting with each other and the environment is clearly missing.

Moreover, we do not require for Cooperating Objects that the coalescence of physical and computational objects have to form a new type of system in which neither part can be distinguished. Although Cooperating Objects need to interact with the physical world they mostly can be seen as separate systems with a clear interface to the environment with specific characteristics.

3.1.2 Internet of Things

The origin of the term "Internet of Things" is commonly attributed to Kevin Ashton who used it in a presentation at Procter & Gamble in 1999. In [4], Ashton explains his original idea on the concept: Computers depend on information created manually by humans, but it would be way more efficient and less error-prone if computers gathered data about things in the real world automatically. Therefore, these things need to be identifiable and sensor technology is needed to observe their status.

Since then, the concept has evolved. In a 2008 conference report of the National Intelligence Council, the phrase Internet of Things "refer[s] to the general idea of things, especially everyday objects, that are readable, recognisable, locatable, addressable, and/or controllable via the Internet" [5]. Compared to the original idea, a two-way approach is now included, i.e. the possibility to address and control an object.

The European Research Cluster on the Internet of Things gives the following definition in its 2011 Strategic Research Roadmap [6]:

> Internet of Things (IoT) [...] could be conceptually defined as a dynamic global network infrastructure with self configuring capabilities based on standard and interoperable communication protocols where physical and virtual 'things' have identities, physical attributes, and virtual personalities, use intelligent interfaces, and are seamlessly integrated into the information network.

Compared to the first two definitions where relatively dumb things are monitored and controlled by an intelligent and global, yet static infrastructure, this hard separation does not hold any longer. Now, smarter, autonomous objects form a dynamic network. This is similar to Cooperating Objects, but we do not require integration on a global scale. Therefore, we also focus on Cooperating Objects that are smaller but more specialised. Although based on standard communication protocols, which is also an important goal for the Cooperating Objects field, tailored higher level interactions will allow a better cooperative behaviour than general and highly interoperable service interfaces. Nevertheless, we are also dealing with the integration of Cooperating Objects into larger scale systems, be it a local IT business infrastructure or a global internetwork.

Interestingly enough, the definition does not argue convincingly for the need of global integration, although it is part of the definition. The Internet of Things can be seen, therefore, as an enabling technology that services can use to interact with these enhanced objects. In that sense, the Internet of Things could also be regarded as foundation or enabling and complimentary technology for Cooperating Objects.

However, the roadmap also elaborates on possible applications: "The major objectives for IoT are the creation of smart environments/spaces and self-aware things (for example: smart transport, products, cities, buildings, rural areas, energy, health, living, etc.) for climate, food, energy, mobility, digital society and health applications." To a large part, these examples overlap with our ideas for Cooperating Objects.

The path to these applications is different. In the Internet of Things, physical things have a virtual representation that is enriched with context data. This is one possible way to represent information gathered from reality. For Cooperating Objects we focus on smart objects that form the cooperative partners in an interaction and that treat their observations as data. Very often, the uniqueness of the observed things can be locally limited so that it is not necessary that these are uniquely identifiable and addressable.

Different views have emerged the last years such as the "web of things" which focuses web technology interaction among the "things"; however we consider that these are covered in the broader definition of Internet of Things.

3.1.3 Machine-to-Machine (M2M)

According to [7] "[t]he term M2M refers to systems that enable machines to communicate with back-end information systems and/or directly with other machines, in order to provide real-time data." The M2M API constitutes a service end point in an Internet of Things architecture and is, therefore, regarded as a key enabler. However, Machine-to-Machine communication often occurs in IT systems of larger scale and with bigger machines that are usually not considered as "everyday objects". Thus, there is a large overlap between the M2M and the IoT area, but M2M is not a complete part of IoT [5]. The same arguments hold for the relation between Cooperating Objects and M2M communication.

Typically, M2M communication is based on open and standardised protocols to minimise the effort for companies when deploying the machines. However, a large number of M2M protocols are in use, e.g. SCADA, UPnP or CoAP, each of which is tailored to a specific use case. We do not aim at creating a universal standard for communication between our smart objects but use the best fitting method for a specific Cooperating Objects application which, in the most complex case, could be composed of different and incompatible standards used through gateways that implement two or more protocols.

3.1.4 Radio-Frequency IDentification (RFID)

The technology used for Radio-Frequency IDentification dates back to the 1940s but was first seen as a method to communicate passively, e.g. [8]. In 1970, Mario

Cardullo filed a patent for the first passive, read-write RFID tag he designed mainly to identify cars [9].

As the name indicates, RFID is mostly used to identify objects, e.g. goods using the Electronic Product Code (EPC), to recognise locations, or to read stored information. Compared to optical systems like (2D-)barcodes, it has the advantage that it does not require line-of-sight and that often the stored information can be changed. RFID applications include supply chain, ticketing, asset tracking, maintenance, retail, and personal identification. Additional global network infrastructure is added to RFID to make the information easily accessible via email or the web [10]. These look very similar to the Internet of Things vision. Not surprisingly, RFID is, therefore, seen as an enabling technology for the Internet of Things.

Although RFID plays also a role in Cooperating Objects the latter focuses more on smart objects that take an active role in the cooperation themselves. Each individual entity of a Cooperating Object is able by design to perform complex processing tasks, if so needed. On the other hand, passive RFID does not perform any kind of processing and only returns an identification as an answer to an external stimulus (the reader). In this sense, the intelligence of a system based on passive RFID technologies lies in the infrastructure and in the readers, but not on the distributed and embedded devices that form the bulk of the network. Active RFID tags feature their own power source e.g. battery or an external source, and are more advanced coming very close to the Cooperating Objects domain, but due to their cost and size their use is comparatively tiny.

3.1.5 Robotics

Based on the 1980s definition of robotics, which is "the intelligent connection between perception and action" [11], a robot could be the perfect incarnation of a Cooperating Object. However, sensing, actuation and computing devices are incorporated in a single machine and, therefore, no real cooperation between autonomous entities is involved in the simplest case. Also, telerobotics cannot be considered Cooperating Objects in the sense of our definition since the robot is controlled remotely and no cooperation with other robots or systems is required. However, even if multiple autonomous robots are involved in a system, this does not make it automatically a Cooperating Object. This is only the case where a number of heterogeneous devices are involved in the overall system. Therefore, important problems like distributed mission and motion planning are still considered part of the core area of robotics.

There are good examples for real involvement of robots in Cooperating Objects, which are mainly based on the cooperation between robots and sensor networks. Sensor networks provide a more complete sensor coverage than a few robots and allow for (better) communication between robots over a multi-hop network. For example, in Smart Homes or in Search and Rescue Scenarios sensor networks deliver context and position information. On the other hand, robots can deploy, maintain and repair sensor networks, as can be seen in several past and currently ongoing projects [12–14].

3.1.6 Internet of Services

The Internet of Services (IoS) is building over the existing Internet by advocating a service-based interaction among all partners. These would be based on open standards and technologies that would enable information exchange and rapid build up of interactions e.g. cooperation. The main target is business interactions among the various stakeholders and promote information reuse/repackaging and integration in order to support business ecosystems and empower the "Internet of People" i.e. the business interactions among people and their organisations. To achieve the objectives a service delivery framework (SDF) must be in place providing the governing rules.

The Internet of Services envisions service-based interactions and this could hold true also for the physical world, where various objects directly or indirectly can offer their functionality as a service [15, 16]. SOA-ready devices blur the borders of Internet of Services and Internet of Things. When it comes down to cooperation, SOA-based cooperation may overcome the problems of heterogeneity (both in IoT, IoS and their combination). Cooperating Objects could engage in cross-layer interactions (as depicted in Fig. 3.4) and hence be an enabling technology for IoS, especially when it comes down to monitoring and control in the physical world. This is increasingly important when considering the emerging cyber-physical infrastructures such as that of the SmartGrid (as depicted in Fig. 3.2) as well as Factory of the Future, where the service-enabled Cooperating Objects are seen as a promising approach [17, 15].

3.1.7 Software Agents

In computer science the agents are software programs acting on behalf of a user or another entity. There are many categories of agents e.g. intelligent agents (which implies capabilities for learning/reasoning), autonomous agents, distributed agents, multi-agent systems, mobile agents etc. Agents are considered one of the most important paradigms for conceptualising, designing, implementing and simulating software systems.

Especially multi-agent systems (MAS) [18] support group behaviour of agents in dynamic situations, and are capable of simulating systems with large number of heterogeneous entities behaving differently. As such MAS are more suitable for evaluating distributed systems that involve complex interaction between entities, e.g. humans, industrial robots, smart devices. As agents are autonomous and operate without human intervention MAS can model really complex non-deterministic systems governed by common and possibly even conflicting goals.

Many of these characteristics already researched in the agent domain, are characteristics that Cooperating Objects will depict in the effort to build up cooperations among them as well as between groups. Hence the agent domain shares common ground with Cooperating Objects and the way they might interact, although the physical constraints play also a key role in the potential behaviour of a Cooperating Object.

3.1.8 Autonomic Systems

Autonomic computing [19] was introduced by IBM as a mean to target increasing computer system complexity and aimed initially at automating management of enterprise computational systems. In "The Vision of Autonomic Computing" [20] it is stated that the dream of interconnectivity of computing systems and devices could become the "nightmare of pervasive computing" in which architects are unable to anticipate, design and maintain the complexity of interactions. The essence of autonomic computing is system self-management, freeing administrators of low-level task management whilst delivering an optimised system. This will get a new dimension once the billions of anticipated devices in the Internet of Things vision are on-line and start cooperating.

Autonomic concepts are very much applicable to Cooperating Objects and are already present in similar areas such as the software agents. We consider that in a highly heterogeneous Cooperating Objects infrastructure autonomic concepts applied at several layers including networking and communication as well as control and management may assist in taming the emerging challenges. For instance in a self-managed autonomic system, the enterprise service does not control the system directly, but only defines general policies and rules that serve as an input for the self-management process. In such complex environments self-* features will be probably evident not only in the enterprise services themselves, but also in the Cooperating Object ecosystem as well as the layers that reside between them. Desirable autonomic features include self-configuration (automatic configuration of components), self-healing (automatic discovery and correction of faults), self-optimisation (automatic monitoring and control of resources to ensure the optimal functioning with respect to the defined requirements) and self-protection (proactive identification and protection from arbitrary attacks).

3.1.9 System of Systems

In the pursue of large scale complex problems, various systems will need to be brought together in order to tackle them. This integration of various heterogeneous independent and self-contained systems in order to solve a global (often multidisciplinary) problem constitutes the emerging domain of System of Systems (SoS).

According to Maier [21] five principal characteristics are useful in distinguishing very large and complex but monolithic systems from true systems-of-systems:

1. Operational Independence of the Elements: If the system-of-systems is disassembled into its component systems the component systems must be able to usefully

operate independently. The system-of-systems is composed of systems which are independent and useful in their own right.
2. Managerial Independence of the Elements: The component systems not only can operate independently, they do operate independently. The component systems are separately acquired and integrated but maintain a continuing operational existence independent of the system-of-systems.
3. Evolutionary Development: The system-of-systems does not appear fully formed. Its development and existence is evolutionary with functions and purposes added, removed, and modified with experience.
4. Emergent Behaviour: The system performs functions and carries out purposes that do not reside in any component system. These behaviours are emergent properties of the entire system-of-systems and cannot be localised to any component system. The principal purposes of the systems-of-systems are fulfilled by these behaviours.
5. Geographic Distribution: The geographic extent of the component systems is large. Large is a nebulous and relative concept as communication capabilities increase, but at a minimum it means that the components can readily exchange only information and not substantial quantities of mass or energy.

In addition to these five criteria by Maier [21], inter-disciplinary study, heterogeneity of systems and networks of systems are also considered characteristics of SoS [22].

The System of Systems domain clearly shares common ground with Cooperating Objects. Today Cooperating Objects are still in embryonic stage, and mostly individual interactions among them are investigated. However in the future we expect that the focus will also include their behaviour as an ecosystem driven by cooperation. Additionally the interactions among such systems (that heavily rely on Cooperating Objects) is an angle whose impact will still need to be investigated.

3.1.10 Summary

Figure 3.1 shows a pictorial representation of the different areas we have just defined taking into account the research landscape around them. As can be seen, there is a clear overlap between all of them, which is the reason why this chapter has tried to shed light on the not so clear differences among them. Although such kind of pictures are always prone to errors and to oversimplify the complexity of each area, there are two aspects that are worth noting:

1. All of the research areas discussed have a strong overlap, with the biggest one showing the very close relationship between Cooperating Objects and Cyber-Physical Systems.
2. Each individual area looks at aspects of a very complex problem from a different and, in most cases, complementary way. For this reason, each area in itself has its raison-d'être although it is also clear that the world will eventually converge

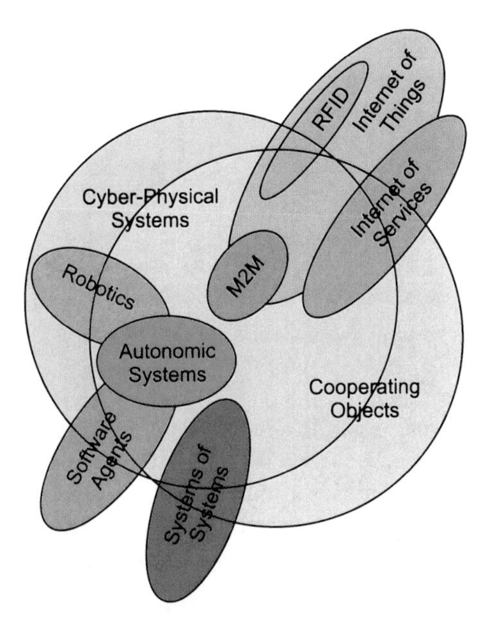

Fig. 3.1 Overview of the different technologies

and, down the road, will consider all of these technologies siblings of each other, if not equivalent.

These two observations highlight the clear need for interdisciplinary research and the fact that researchers of each individual area need to be aware and, in some cases, fully understand the concepts and solutions developed in other neighbouring research areas.

3.2 Key Technology Trends

We are witnessing today several technology trends that may have a significant impact when designing and implementing solutions depending on Cooperating Objects. Especially for critical infrastructures, timely high-quality information acquisition, context assessment, decision-making and actuation needs to be open, streamlined and dependable [23]. Considering its complexity and dynamic interactions, modelling and assessing the tremendous amount of information generated by the future heterogeneous Cooperating Objects dominated infrastructures, is a very challenging task. Integration and cooperation are major goals especially for domains such as the SmartGrids or Industrial Automation that are relative new to IT technologies and their rapid evolution pace. For instance Cooperating Objects may play a key role within the future smart house as well as with its integration in the smart grid [17] as depicted in Sect. 3.3. Furthermore future industrial systems such as SCADA/DCS may heavily depend on Cooperating Objects to deliver the desired functionality [24].

3.2.1 Information Driven Integration

Business systems nowadays heavily depend on high-quality information from the real world. Cooperating Objects might assist by providing that link i.e. dynamically discover, integrate, and interact with the real world with collaboration as a key behaviour. The service oriented architecture (SOA) empowered Cooperating Objects may point us towards a potentially right direction. By abstracting from the actual underlying hardware and communication-driven interaction and focusing on the information available via services, we move towards a service & information driven interaction. By accessing the isolated information and making the relevant correlations, business services could further evolve, and dynamically integrate real-time feedback from the real physical-domain services (both for monitoring and control).

3.2.2 On-Device Business Process Execution

In an Cooperating Objects infrastructure, large amounts of data are usually generated. Its transmission to a central location in order to extract the gist may not be optimal and the enterprise systems are not designed for such operations today. Hence business processes that require data residing on a Cooperating Objects, could outsource that part of their functionality to run directly on the Cooperating Object or a collocated system (depending capabilities and characteristics e.g. communication, computation and possibly spatial constraints). Distributing load in the layers between enterprises and the real world infrastructure (distributed business process) is not the only reason; distributing business intelligence is also a significant motivation.

3.2.3 Cooperation/Coopetition

Most of Cooperating Objects today rely on operating in standalone manner or provide information to standalone services. However with the increased communication and emergence of networks of Cooperating Objects, they will be able to cooperate, share information, act as part of communities and generally be active elements of a more complex system. By doing so they may be able to tackle aspects envisioned for emerging large scale systems such as self-management, self-optimisation, and self-healing. As such the governing logic may be expressed in a goal oriented manner assigned to networks of Cooperating Objects aiming at satisfying business process requirements. Coopetition is also part of Cooperating Objects as they may only partially collaborate on common interest areas or although competing try to commonly seize specific opportunities.

3.2.4 Virtualisation and Utility Computing for Cooperating Objects

In the IT world we witness a trend towards virtualisation of resources such as hardware platforms, operating systems, storage devices, network resources etc. Virtualisation addresses many enterprise needs for scalability, more efficient use of resources, and lower Total Cost of Ownership (TCO) just to name a few. Cooperating Objects may be directly affected by this trend as now functionality provided by a single standalone Cooperating Object can "roam" the network of Cooperating Objects empowered devices, to optimally execute depending on abstract provided resources residing both on the physical world (e.g. available sensory instruments) as well as the IT world (e.g. cloud available computing power). Important is also that although the Cooperating Object may not have the capabilities required by itself, it can extend these by collaborating/outsourcing computational aspects to services in the cloud and possibly physical tasks to nearby Cooperating Objects equipped with the required capabilities e.g. sensors/actuators.

3.2.5 Multi-Core and GPU Computing for Cooperating Objects

Assessment of the huge amounts of data generated by Cooperating Objects will be challenging and require significant processing power. Since 2005 we have seen the emergence of multi-core systems, that nowadays exist also in constrained devices e.g. dual-core powered android smartphones. The general trend is towards chips with tens or even hundreds of cores. Advanced features such as simultaneous multithreading, memory-on-chip, etc. promise high performance and a new generation of parallel applications unseen before in embedded CPS. Additionally in the last decade we have seen the emergence of GPU computing where computer graphic cards are taking advantage of their massive floating-point computational power to do stream processing.

For certain applications this may mean a performance increase to several orders of magnitude when compared with a conventional CPU. Furthermore a recent trend of integrating built-in graphics capabilities with processors (graphics-enabled microprocessors—GEM) like Intel's Sandy Bridge and AMD's Fusion, may imply that capabilities of GPU computing may be available to any kind of device hosting one of those processors. Such a CPU/GPU hybrid can possibly be even more efficient by removing the slow communication between CPU and GPU. The processors with built-in graphics capabilities to be installed in 2011 on 115 million notebooks account for half of total shipments. By 2014, 83% of the world's notebooks and 76% of desktops will ship with graphics-enabled microprocessors [25]. This in practice implies that CPS equipped with such technologies may increasingly possess significant computational power, be more energy efficient, and possibly execute high performance stream processing e.g. on metering data at the point of action.

3.2.6 SOA-Ready Cooperating Objects

The future infrastructure will be ubiquitous and highly heterogeneous, and so will be the majority of Cooperating Objects they will rely upon. One of the trends e.g. in Industrial Automation is to enable Cooperating Objects to offer their functionality as one or more services for consumption by other entities. Due to these advances we are slowly witnessing a paradigm shift where devices can offer more advanced access to their functionality and even host and execute business intelligence, therefore effectively providing the building blocks for expansion of service-oriented architecture concepts down to the device layer [15]. This could greatly enhance interoperability and lead to independent layered development at software (potentially also to hardware) side. Furthermore the integration is done via the capabilities offered (as services), while the actual implementation is hidden.

3.3 Applications

There are many applications areas for Cooperating Objects. We just indicatively mention some of them here in order to give an taste of what Cooperating Objects can do in some hot research areas today.

3.3.1 Internet of Energy

In the future service-based Internet of Energy [26], several alternative energy providers, like legacy providers and households, are interconnected. Via "smart meters", one is able to interact with a service based infrastructure and perform actions such as selling and buying electricity independently. Both the smart meters as well as the household appliances are envisioned to be Cooperating Objects and participate in complex interactions. More advanced services are envisioned that will take advantage of the near real-time information flows among all participants. Furthermore the energy consuming/producing devices will be no more considered as black-boxes but will also get interconnected, which will provide fine-grained info e.g. energy optimisation per device. Existing efforts in the emerging Internet of Things and Internet of Services, will be combined and be a crucial part in the envisioned Internet of Energy (Fig. 3.2).

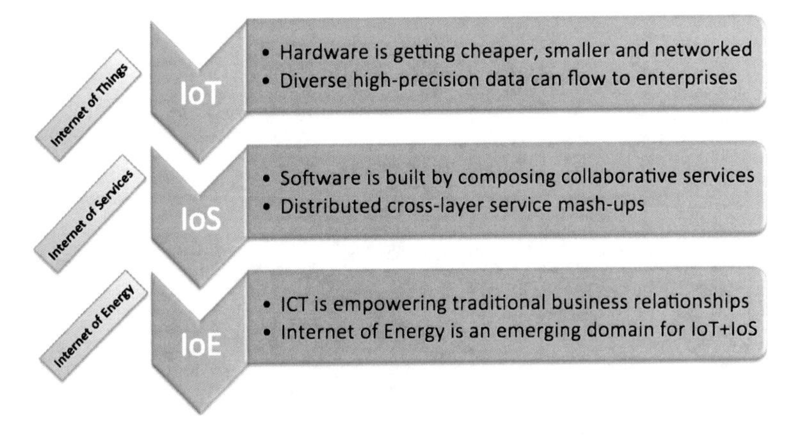

Fig. 3.2 Internet of energy: combining IoT and IoS

The strong coupling of Information and Communication (ICT) technologies—especially via the usage of networked embedded devices—with the energy domain, is leading to a sophisticated dynamic ecosystem referred to as the Internet of Energy. In the last mile of the Smart Grid i.e. the future smart home, heterogeneous devices will be able to measure and share their energy consumption, and actively cooperate with house-wide or building wide energy management systems as well as external services. The emerging Smart Grid will heavily depend on cooperation that will emerge at various layers (horizontally and vertically), and on the interaction with networked embedded systems i.e. Cooperating Objects that will be realising also its sensing and actuation functionality.

The SmartGrid is a critical infrastructure and should be robust, self-managed, self-sustained and enable dynamic reorganisation and coordination of services and markets; the Internet-based infrastructure should be tightly linked to the energy domain, and be used to support the development of new mechanisms for coordinating real-time demand and responses in the electricity market—i.e. the consumption and feeding-in of power and the resulting interlinked commercial transactions. Transaction platforms will serve as electronic marketplaces facilitating the commercial activity associated with the purchase and sale of electricity and its derivatives, not only for utility companies but also for decentralised consumers and producers. Intelligent, interactive energy-management systems will be needed for an infrastructure capable of supporting the deregulated energy market.

Treating homes, offices and commercial buildings as intelligently networked cooperations (driven by Cooperating Objects) can contribute towards enhancing the efficient use of energy. When smart houses are able to communicate, interact and negotiate with both customers and energy devices in the local grid, the energy consumption can be better adapted to the available energy supply, especially when the proportion of variable renewable generation is high. Several efforts focus on integrating the smart houses and the emerging smart grids.

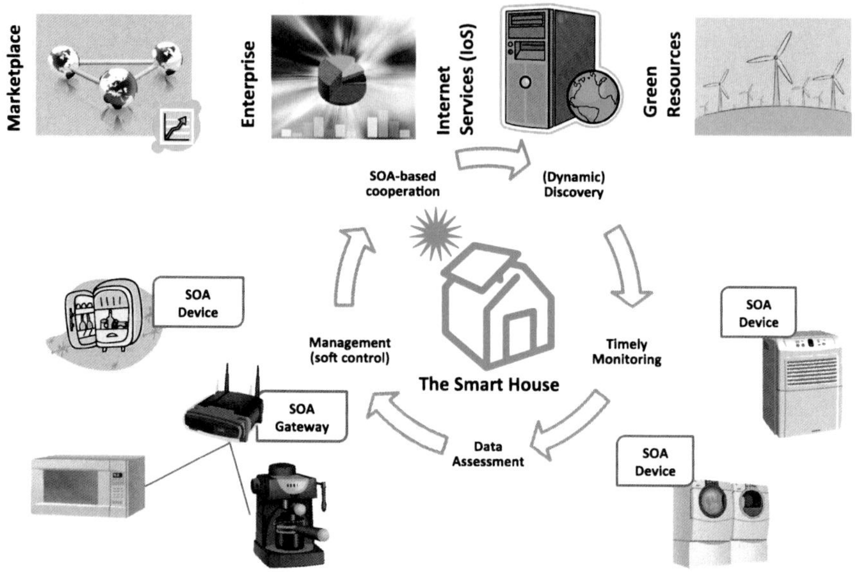

Fig. 3.3 Collaboration within the smart house and with external entities

In the context of Smart Grid, several entities can fall within the context drawn by Cooperating Objects. Typical examples are advanced smart meters, smart white label appliances, electric cars, various prosumer/consumption/production devices, alternative energy resources, etc. In typical cooperating object demonstrators, all of these are capable of providing their functionality (e.g. energy consumption, status, management etc.) as a service that can be utilised to achieve better energy management in standalone mode or as part of more complex system.

3.3.2 Smart House/City/Planet

The smart house of the future [17] will be able to collaborate with numerous external entities, let it be alternative energy resources, marketplaces, enterprises, energy providers etc. The de facto standard for high-level communication today is via (web) services, which allows for flexible functionality integration without revealing the details of the actual implementation. Therefore the heterogeneity is hidden, while a common service-based interaction (Fig. 3.3) is empowering the creation of sophisticated applications. As such the smart house will be part of a complex system of systems.

Apart from the out-of-smart house interactions, the collaboration will be also visible within the house itself. We already have numerous protocols and even different technologies at hardware and communication layer, which inevitably will increase in the future. It is however a common belief that all of this heterogeneity will be hidden behind gateways and (service) mediators, which will eventually allow the device to tap into an IP-based infrastructure, using therefore Internet standards. Already today the IP protocol is developed further to run in tiny and resource constrained devices, while with the IPv6 (and 6LoWPAN) any device will have its own IP address and therefore be directly addressable (and possibly uniquely identifiable).

Due to IP penetration down to discrete device level, it is expected that devices will not only provide their information for monitoring to controlling entities, but will be able to dynamically discover nearby devices and collaborate with them (as depicted in Fig. 3.3). As such P2P interactions will emerge, which can be exploited by locally running applications that execute monitoring or controlling tasks. It is expected that each appliance manufacturer will make optimisations so that his device operates e.g. as efficient as possible. However it is beyond of current capabilities to see how this device will function collectively in an environment composed of other devices, as this environment is not standard and can not be known a priori. Here the collaboration concepts come in play, and the end-user (or another third party service provider) can create ad-hoc highly customisable applications that take into consideration the local context (e.g. of the specific house) and organise house-wide, building wide or even neighbourhood wide optimisations.

Devices on the smart house are and will remain highly heterogeneous both in hardware and software. As such we need to find a way that this heterogeneity is abstracted and still communication among them (and collaboration) can be achieved. The development of middleware approaches that act as "glue" for device to business connectivity (and later also for device to device connectivity) is a viable approach. However it is more efficient if the heterogeneity is tackled at device level and only a limited part is delegated at the middleware. As such it is preferred that the devices offer standardised interfaces e.g. complying to ZigBee profiles and then limited connectors at the middleware side (for classes of devices and not specific to manufacturer or device) can directly enable their connectivity to on-line services and enterprise systems.

In parallel to local collaboration, devices with advanced capabilities will be able to interact with network-based services hosted in enterprise systems, or simply somewhere on the Internet [15]. These devices will be able to enhance their own functionality in a dynamic way by invoking services that were not thought of at the time of device design. Price signals are often brought up as a key functionality that would affect the device behaviour; for instance a device would get a price signal e.g. from the energy provider (or by monitoring or collaborating with an on-line service) and adapt its functionality.

On city wide level, the different systems are envisioned to cooperate and be goal-driven with consideration of the context they operate in. Tackling especially goals of energy efficiency, global management etc. over a highly complex and heterogeneous infrastructure as that of a city implies cooperation based on both social and environmental directions. Hence multi-disciplinary considerations will need to be made towards smart economy, smart mobility, a smart environment, smart people, smart living, and, finally smart governance. Cooperating Objects are considered as an enabling factor that may assist towards realising and measuring these goals. As an example, wireless sensor networks are already in use for monitoring smart city pollution, traffic jams, parking, water leaks, waste etc. The availability of ICT is a key indicator of a smart city, however equally or even more important is the quality of ICT as well as the interactions they enable, and here Cooperating Objects could play a pivotal role towards future inclusive and sustainable cities.

Today cities struggle with traffic congestion, water management, communication technology, smart grids, healthcare solutions, rail transportation, etc. and therefore IBM has created its Smarter Cities[1] portal, which tracks progress on these issues in several key cities around the world. This is part of IBM's global initiative "Smarter Planet"[2], where they seek to highlight how by advancing all systems in cities and more, to be more instrumented, intelligent and interconnected requires a profound shift in management and governance toward far more collaborative approaches. The "Smarter Planet" initiative identifies six fundamental needs related to:

- Business Analytics and Optimisation
- Smarter Computing
- Business Agility
- Social Business
- Smarter Products and Services and
- Security and Resilience.

In this approach cooperation is the key aspect, and therefore Cooperating Objects are seen as the backbone of it.

3.3.3 Factory of the Future

The last decade has witnessed a deep paradigm shift on the shop-floor where Information and Communication Technologies (ICT) are being used extensively. As high performance micro-controllers are being embedded in devices used in manufacturing and process automation, services hosted on them will enable new applications that could significantly increase the efficiency of current shop floor systems [27]. Today, integrating devices in applications requires not only advanced knowledge of the device, its configuration and the way it connects, but also the installation of highly specialised software that glues the data (often in proprietary format) with applications. Such an integration model is costly, application specific and create isolated islands for each shop floor. As a result, it is extremely hard for enterprise service developers to enrich service functionality with real time data coming from the shop floor. The SOA concept, has proven very successful for gluing heterogeneous systems, and if the same would be applicable for devices this would be a significant step forward in the direction of coupling the real-world and the business world.

---

[1] http://www.ibm.com/smartercities

[2] http://www.ibm.com/smarterplanet/now

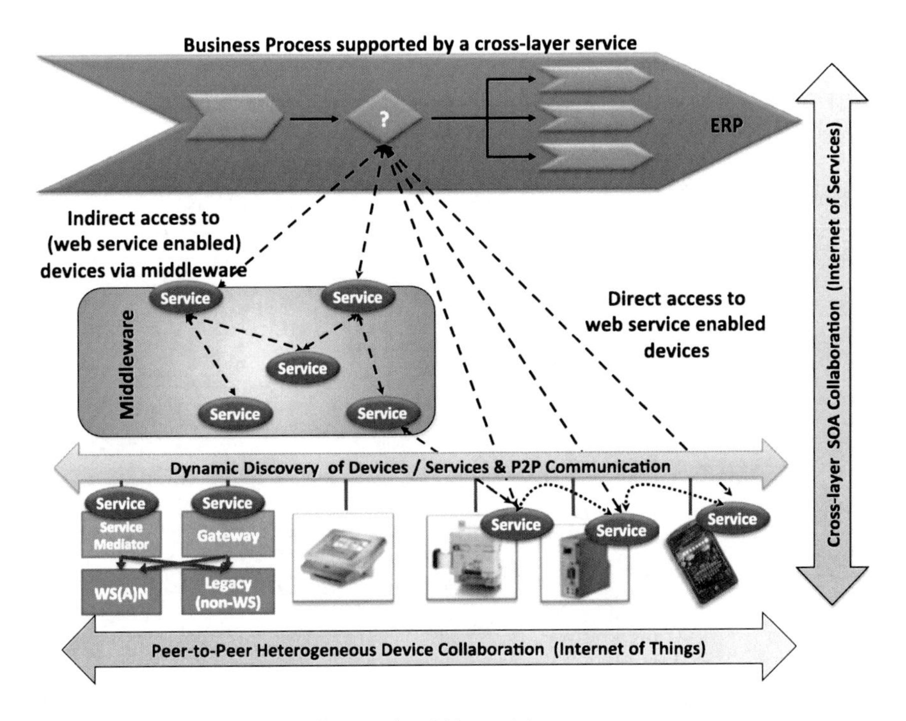

Fig. 3.4 Horizontal and vertical Cooperating Objects vision

Future shop floor infrastructures can significantly benefit from service-oriented approaches, both in vertical (cross-level) and horizontal communication, as for instance demonstrated within the SOCRADES project (www.socrades.eu). In these infrastructures, new, rich services can be created by orchestrating and combining services from different system levels [15], i.e. services provided by enterprise systems, by middleware systems on the network, and by devices themselves (as depicted in Fig. 3.4). The composed services with complex behaviour can be created at any layer (even at device layer). In parallel, dynamic discovery and peer-to-peer (P2P) communication allows for dynamically discovery of functionality of each device. The trend is to clearly move away from proprietary connections between monolithic hardware and software systems towards more autonomous systems that interact in a more standardised, cooperative and open way.

Focusing on cooperation and taking advantage of the capabilities of Cooperating Objects poses a challenging but also very promising change on the way future factories will operate, as well as to the way we design software and model their interactions. Interaction between the cooperating automation units that are able to expose and/or consume services, for each production scenario in a defined production domain, e.g. electronics assembly, manufacturing, continuous process, etc. may enable a new breed of applications e.g. collaborative manufacturing. Example of cooperating objects in the factory of the future can be a simple intelligent sensor or a part/component of a modular machine, a whole machine or even a complete production system.

The convergence of applications and products towards the SOA paradigm improves shop floor integration and transparency, thereby increasing reactivity and performance of the workflows and business processes commonly found in manufacturing and logistics. Events become available to any entity of the system as they happen, and business-level applications can exploit such timely information for purposes such as diagnostics, performance indications, or traceability. While these vertical collaborations are beneficial for business application software, new challenges arise: direct communication with devices can be error prone or unreliable, which must be considered when critical decisions, such as branches in a work-flow, depend on it.

References

1. Rajkumar R, Lee I (2006) NSF workshop on cyber-physical systems. http://varma.ece.cmu.edu/cps/
2. National Science Foundation (2011) Cyber-physical systems (CPS) program solicitation NSF 11-516. http://www.nsf.gov/funding/pgm_summ.jsp?pims_id=503286
3. CPS Summit Steering Committee (2008) Cyber-physical systems executive summary. http://varma.ece.cmu.edu/Summit/CPS-Executive-Summary.pdf
4. Ashton K (2009) That 'internet of things' thing. RFID J. http://www.rfidjournal.com/article/view/4986
5. SRI Consulting Business Intelligence (2008) Appendix F: the internet of things (background). In: Disruptive civil technologies conference report (CR 2008–07), National Intelligence Council. http://www.dni.gov/nic/confreports_disruptive_tech.html
6. European Research Cluster on the Internet of Things (2011) Internet of things strategic research roadmap. http://www.internet-of-things-research.eu/pdf/IoT_Cluster_Strategic_Research_Agenda_2011.pdf
7. Internet-of-Things Architecture (IoT-A) Consortium (2011) Project deliverable D3.1—initial M2M API analysis. http://www.iot-a.eu/public/public-documents/documents-1/1/1/d3.1/
8. Stockman H (1948) Communication by means of reflected power. Proc Inst Radio Eng 36(10):1196–1204. doi:10.1109/JRPROC.1948.226245
9. Cardullo M (2003) Genesis of the versatile RFID tag. RFID J. http://www.rfidjournal.com/article/view/392
10. Roussos G, Kostakos V (2009) RFID in pervasive computing: state-of-the-art and outlook. Pervasive Mobile Comput 5:110–131. doi:10.1016/j.pmcj.2008.11.004
11. Siciliano B, Khatib O (eds) (2008) Springer handbook of robotics. Springer, Berlin
12. AWARE Consortium (2011) AWARE project webpage. http://grvc.us.es/aware/
13. EC-SAFEMOBIL Consortium (2011) EC-SAFEMOBIL project webpage. http://ec-safemobil-project.eu
14. PLANET Consortium (2011) PLANET project webpage. http://www.planet-ict.eu
15. Karnouskos S, Savio D, Spiess P, Guinard D, Trifa V, Baecker O (2010) Real world service interaction with enterprise systems in dynamic manufacturing environments. In: Benyoucef L, Grabot B (eds) Artificial intelligence techniques for networked manufacturing enterprises management. Springer, ISBN 978-1-84996-118-9
16. Spiess P, Karnouskos S, Guinard D, Savio D, Baecker O, Souza LMSD, Trifa V (2009) Soa-based integration of the internet of things in enterprise services. In: IEEE international conference on web services, ICWS 2009, Los Angeles, CA, USA, pp 968–975. doi:10.1109/ICWS.2009.98

17. Karnouskos S (2010) The cooperative internet of things enabled smart grid. In: Proceedings of the 14th IEEE international symposium on consumer electronics (ISCE2010), Braunschweig, Germany, 07 June 2010
18. Wooldridge MJ (2009) An introduction to multiagent systems. Wiley, Chichester
19. Sterritt R, Parashar M, Tianfield H, Unland R (2005) A concise introduction to autonomic computing. Adv Eng Inform 19(3):181–187. doi:10.1016/j.aei.2005.05.012
20. Kephart JO, Chess DM (2003) The vision of autonomic computing. IEEE Comput 36(1):41–50
21. Maier MW (1998) Architecting principles for systems-of-systems. Syst Eng 1(4):267–284. doi:10.1002/(SICI)1520-6858(1998)1:4<267::AID-SYS3>3.0.CO;2-D
22. DeLaurentis DA (2005) Understanding transportation as a system-of-systems design problem. In: 43rd AIAA aerospace sciences meeting, Reno NV, American Institute of Aeronautics and Astronautics, pp 1–14. http://pdf.aiaa.org/preview/CDReadyMASM05_666/PV2005_123.pdf
23. Karnouskos S (2011) Cyber-physical systems in the SmartGrid. In: IEEE 9th international conference on industrial informatics (INDIN), Lisbon, Portugal
24. Karnouskos S, Colombo AW (2011) Architecting the next generation of service-based SCADA/DCS system of systems. In: 37th annual conference of the IEEE industrial electronics society (IECON 2011), Melbourne, Australia
25. Jennings R (2011) Analyst: nearly half of all PCs to use graphics processors. http://www.techworld.com.au/article/380121/analyst_nearly_half_all_pcs_use_graphics_processors/
26. Karnouskos S, Terzidis O (2007) Towards an information infrastructure for the future internet of energy. In: Kommunikation in Verteilten Systemen, KiVS (2007) conference, Bern, Switzerland. VDE Verlag, ISBN 978-3-8007-2980-7
27. Colombo AW, Karnouskos S (2009) Towards the factory of the future: a service-oriented cross-layer infrastructure. In: ETSI (ed) ICT shaping the world: a scientific view, vol 65–81. Wiley, New York

Chapter 4
Enabling Technologies

Nowadays, it is impossible to create or work on technologies that do not rely more or less heavily on the development of other areas. New developments in these related areas usually go hand-in-hand, and a major breakthrough in one of the enabling technologies can really boost the work that can be performed on the other areas.

This is also true for Cooperating Objects and, as we have seen in the previous sections, Cooperating Objects have emerged as a combination and natural extension of already existing research areas that have been evolving rapidly in the past years. Therefore, it is worth pointing out more precisely what we consider are the major pillars for research in Cooperating Objects.

4.1 Hardware

The hardware platforms of Cooperating Objects are as diverse as are the single areas of which Cooperating Objects consist. Thus, there is no single architecture—and probably will never be since the heterogeneity of the devices is a strength of Cooperating Objects. Nevertheless, some common issues can be identified in the hardware sector.

4.1.1 Energy

Since Cooperating Objects often consist of mobile devices the continuous availability of energy is an important issue. The storage of energy, e.g. in batteries, is limited, especially when form factors have to be met. Power generation using ambient energy sources such as vibrations, motion or heat is still in its infancy. But even when energy seems to be available, for example in cars, the increasing number of electrical devices depending on one power source can become a problem. Therefore, beside energy

P. J. Marrón et al., *The Emerging Domain of Cooperating Objects*,
SpringerBriefs in Cooperating Objects, DOI: 10.1007/978-3-642-28469-4_4,

harvesting, the single devices of a Cooperating Object need the be very energy efficient, both hardware and software.

4.1.2 Sensors and Actuators

Sensors and actuators are important since they enable the tight interaction of Cooperating Objects with the environment and provide context awareness. They are as diverse as the hardware platforms and the applications. Challenging tasks are the trade-off between low-cost sensors vs. sensor capability, i.e. several cheap sensors can be used instead of a single expensive one, the usage and combination of different sensor types measuring the same physical phenomenon, or the (re)calibration of sensors in a dynamic Cooperating Objects scenario.

4.1.3 Communication Hardware

The easiest way to enable cooperation is by direct communication (see Sect. 2.6). However, promising Cooperating Objects are hindered since common communication interfaces are missing, for example mobile phone with 802.15.4 interfaces, which are often used in Wireless Sensor Networks, are rarely to be found. Antenna design for small devices is also a challenging task. Also, other means of communication beside radio could be used in special environments, e.g. under water.

4.2 Algorithms

To run a Cooperating Object, several basic algorithms need to be available that support the actual goal. Some algorithms are specific to one device type, e.g. motion planning for the involved robots, but other algorithms are of general use for all devices in a Cooperating Object.

4.2.1 Communication Protocols

Having compatible hardware is only one part to the solution, but the software protocols need to be compatible and suitable for the devices as well. A lot of research has been done mainly in the areas MAC and routing. However, further improvement is needed in terms of power consumption and to support highly mobile scenarios.

Non-functional properties such as timeliness or reliability are gaining importance. Bandwidth estimation and admission control in networks with dynamic channel conditions and dynamic devices as well as radio resource management to adapt transmission frequencies and power, channel access timings, coding or modulation are needed to meet these new demands.

4.2.2 Time Synchronisation

A common time base is essential for many tasks in Cooperating Objects, for example the synchronisation of wake up periods, the correlation of real-world observations and coordinated actions. Time synchronisation protocols that try to achieve this common time base have already reached a mature state. To improve the reliability, data-driven synchronisation, which relies on observed events instead of direct communication, deterministic bounds on the synchronisation error and secure synchronisation mechanisms are investigated.

4.2.3 Localisation

Location and orientation of objects is already valuable information in many areas like robotics or Wireless Sensor Networks and, of course, remains important when moving to Cooperating Objects. It is used from low-level services such as routing to access devices in a certain geographic region to application level, for example, to enable tracking. Many techniques are used to measure or estimate location and orientation, but still accurate, cheap and in-door solutions are an active research field.

4.2.4 Data Processing

Depending on the type of application large amount of data is being generated inside a Cooperating Object. Often, this data is not of immediate or detailed interest for a user so that it needs to be processed, stored and queried inside the network. Several techniques have been explored, mainly concerning Wireless Sensor Networks or peer-to-peer systems, but they often disregard the heterogeneity of Cooperating Objects by abstracting from the device level.

4.2.5 Cooperation Mechanisms

Since a certain level of cooperation is needed for every Cooperating Object this is a key technology. It includes negotiation and decision about goals and sub-goals, planning to achieve the goals, distribution of tasks or roles to the involved devices.

Specialised algorithms, e.g. path planning for robotics, have been developed, but general approaches to meet the various requirements of Cooperating Objects are a big challenge.

4.3 Non-Functional Properties

Properties of a system that affect its quality but not its functionality are called non-functional properties. In contrast, the non-function properties affect all functional areas. Compared to traditional Quality of Service criteria the perspective in Cooperating Objects needs to be extended to include unique features of this research area.

4.3.1 Scalability

Usually, the term scalability refers to the overall number of devices, which is also correct for Cooperating Objects since the number of involved devices can vary from a few to thousands. But it can also be seen as spatial density or the dimension of the geographical region under coverage since Cooperating Objects are always connected to the real-world.

While the efforts for the single tasks could be reduced when distributing the work to a larger number of participants, the coordination effort increases. In a dense network, communication via a shared wireless channel becomes more difficult, but gathered information can be more fine-grained, but also redundant. All these factors have to be taken into account when designing technologies for Cooperating Objects.

4.3.2 Timeliness

Timeliness represents the timing behaviour of a system with respect to computation and communication and includes issues such as message transmission delay, task execution time, task and message priority, etc. It becomes important when Cooperating Objects are involved in or constitute control loops with sensing and actuating real-world objects where time bounds have to be met. But also mission critical monitoring applications expose time bounds although they might not be as tight. Since in traditional real-time systems the goals are met by over-allocation of resources this approach can usually not be taken for Cooperating Objects due to the energy efficiency goal.

4.3.3 Reliability and Robustness

A Cooperating Objects must perform its required functionality not only under predefined conditions for a specified period of time (reliability) but also under abnormal conditions that have not been foreseen (robustness). Both requirements arise from the fact that Cooperating Objects often have to perform critical operations but are subject to often harsh environmental conditions. Thus, these properties affect both hardware and software to prevent but also detect and solve faults.

4.3.4 Mobility

Due to the embedding in the real-world at least some devices are likely to change their positions, either because they are moved or because the Cooperating Object decides to move a device using actuators. In both cases, the characteristics of the Cooperating Object will change, for example the communication links, the sensor coverage or the unavailability of resources. In case of passive movement, the Cooperating Object has to cope with the changing conditions while active movement can be used to increase the performance of the Cooperating Object.

4.3.5 Security and Privacy

When several devices create a network in an ad-hoc manner it is easy for a malicious device to be part of this network. But even with a fixed set of devices, the use of a wireless communication channel enables an attacker to inject and modify packets. Security techniques like encryption are therefore needed, but require special attention due to the different capabilities of the devices involved in a Cooperating Object, their dynamic nature and the absence of a trusted, central authority.

Privacy issues arise from the fact that events in the real-world are monitored that can expose private data. Therefore, this data must be available to trusted parties only. But even inside a secured network, sensitive data must be restricted and shared in a controlled fashion. If data must not be linked to real persons it is necessary to anonymise or blurring the data and to hide traffic patterns.

4.3.6 Heterogeneity

Heterogeneity is an inherent property of Cooperating Objects (see Sect. 2.3) and both a feature and a problem. As explained earlier, it is often essential to achieve the common goal. On the other hand, different computational capabilities, communication hardware, operating systems, etc. need to be considered when creating Cooperating Objects applications. Thus, the challenge is not only to mitigate the issues arising from the differences between the single devices but at the same time to use them advantageously.

Since a device can also be part of multiple Cooperating Objects and of different applications the requirements to a Cooperating Object and the devices can be heterogeneous as well. This type of heterogeneity need to by tackled in a different way, e.g. by adaptivity and of course through the autonomy of each device to participate in a Cooperating Object.

4.4 Systems

In this section, we focus on technologies that support the development of Cooperating Objects. Although Operating Systems provide basic functionality, e.g. hardware access or memory management, we do not consider it a special enabling technology for Cooperating Objects since they are a foundation for practically every computing device.

4.4.1 Programming Abstractions

In general, programming abstractions and middleware try to hide the complexity of the underlying platforms or systems or to provide additional functionality compared to the operating system. Without them, creating a bigger Cooperating Object application would be a very difficult task. The approaches include: virtual machines; abstractions from nodes, resources or the whole network; role assignment, task definition and distribution; frameworks to enable adaptation, cross-layer interactions or data sharing; context management frameworks.

It is very important the these programming abstractions especially support the non-functional properties that have been describe in Sect. 4.3 since they span all functional aspects of a Cooperating Object and, therefore, need to be tackled in a holistic way.

4.4.2 Debugging, Testing and Verification

Diagnosing faulty behaviour or insufficient performance in Cooperating Objects as potentially large, distributed, heterogeneous system is an important but difficult task, both at the end of the development process and during normal operation. During development, special instrumentation of the devices itself or additional debugging devices can be used. In contrast, for normal operation self-monitoring and self-healing approaches need to be considered that entail only minimal additional costs.

4.5 Others

This category does not affect the software of Cooperating Objects itself but helps in their development and installation.

4.5.1 Modelling and Planning

Modelling and planning tools help to design the deployment of the system. Analytical models can be developed for many parts of a Cooperating Object, such as communication, energy, sensing and mobility, and are essential both for their simulation and verification. Obviously, the accuracy of the models directly affect the validity of the simulation and verification results.

If parts of a Cooperating Object should be installed statically proper planning is required to ensure the intended functionality, e.g. communication and sensor coverage or performance guarantees. The already mentioned models build the core of this planning process, which can be improved further by test installations.

4.5.2 Testing and Verification

In order to detect possible problems, thorough testing is usually performed during the development process. Initial testing can be done using general testbeds, but more accurate results can be obtained by test installations that resemble the final scenario. Common practises still have to be developed for this process since the complexity of Cooperating Objects usually make a complete test infeasible.

If testing is not possible, the correctness of parts of Cooperating Object functionality, e.g. single algorithms, can be proved using formal verification techniques. This is especially important when Cooperating Objects are employed in critical systems for which proper operation has to be guaranteed.

4.5.3 Standardisation

Standardisation of hardware and software is necessary to enable the interoperability and, thus, the cooperation of many Cooperating Objects. Due to the different fields involved in Cooperating Objects several standards will have to be followed and interfaces between them have to be created.

Otherwise all solutions will remain a separate system and will not profit from each other. Moreover, the vision of "Cooperating Objects Everywhere" will not become true, i.e. that most devices will have the ability to cooperate and are actually participating in various and changing Cooperating Objects.

Chapter 5
Conclusions and Outlook

It is clear that the field of Cooperating Objects is a very dynamic one that has the potential of drastically changing the way people interact with the physical world. We are still at the dawn of an era where the convergence of different technologies and the collaboration and cooperation of existing technologies will lead to unprecedented solutions. We have already started to see the effects of how new trends in technology affect the way people socialise, communicate and, in the long-run, cooperate.

In this first book, we have given a clear definition of Cooperating Objects and have also tried to provide a clear picture of current technologies, showing their similarities and differences with respect to Cooperating Objects. We have also outlined the important research aspects and topics that require more attention and continuous work that stems from the research of groups all over the world.

5.1 Breadth of Topics Covered by the Book Series

The complexity of the field and the fact that it is still in its infancy makes it a very prolific ground for topics that require interdisciplinary research. In general, new developments in the related areas of Cooperating Objects and the Internet of Things go hand-in-hand, and a major breakthrough in one of the enabling technologies can really boost the work that can be performed on the other areas.

Figure 5.1 shows the main topic areas envisioned for the book series. Each one of these topics encompasses other subtopics where specific books can be written. A brief description of the purpose and idea behind each main topic is given below.

General Definitions: General books related to the proper definition of terms as well as their difference to other related topics that might appear in neighbouring fields. These books should serve as a reference and for consensus from the community.

Roadmaps, Future Technologies: Recurring books that deal with hot topic from different perspectives from the community but always looking into the future and

P. J. Marrón et al., *The Emerging Domain of Cooperating Objects*,
SpringerBriefs in Cooperating Objects, DOI: 10.1007/978-3-642-28469-4_5,

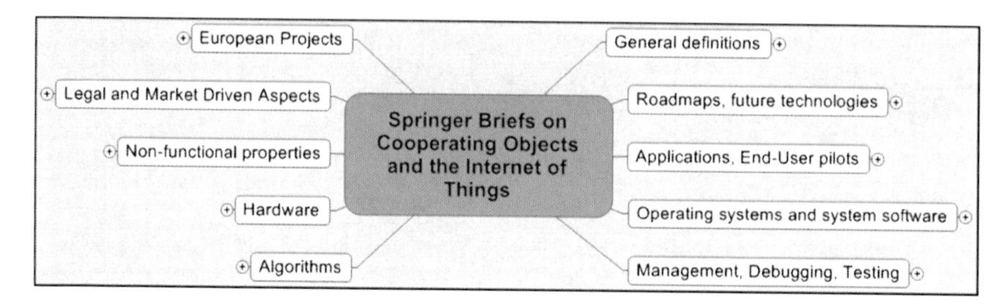

Fig. 5.1 Main topic areas

trying to come up with suggestions for the organisation of research in years to come.

Applications, End-User Pilots: Books that explain in detail specific applications or end-user pilots where experiments and large-scale deployments of the technologies related to Cooperating Objects and the Internet of Things have been performed.

Operating Systems and System Software: Books dealing with topics on the lower layers of the software, from the operating system to middleware systems that provide abstractions to be used by the application developer or the end-user.

Management, Debugging and Testing: Books dealing with topics related to the deployment and maintenance of deployed systems of Cooperating Objects. These topics also include compiler techniques as well as monitoring, debugging and testing techniques.

Algorithms: Books that deal with specific algorithms that enable the processing, localisation, etc. of Cooperating Objects. This is one of the areas that will have a large number of books since there are a number of techniques needed for the proper functioning of Cooperating Objects.

Hardware: Books that deal with hardware issues for Cooperating Objects. This includes the definition of the hardware, new or novel hardware components or architectures as well as energy harvesting techniques that enable the use of Cooperating Object technologies for longer periods of time or even without batteries.

Non-functional Properties: Books that deal with topics such as scalability, quality of service, etc. for one or more types of algorithms.

Legal and Market-Driven Aspects: Books that deal with market research on specific areas of Cooperating Objects or with the legal aspects related with the adoption of technologies by a wider audience.

European Projects: Books that explain the main results of European Projects and try to disseminate the work performed as part of collaborative research in the area.

Our goal is to serve as the series of reference for Cooperating Objects in particular and, more generally, for other related areas and provide a forum for experts in the field to publish their mature results in a well-established forum with high impact and visibility.

5.2 Next Books in the Series

- Research Roadmap for Cooperating Objects
- 22@Barcelona: The Smart City District of Barcelona
- The Emerging Domain of Cooperating Objects: Applications and Markets